Python for Beginners

A Crash Course Guide to Learn Python in 1 Week

by

Timothy C. Needham

Copyright 2017 - All Rights Reserved

Timothy C. Needham

ALL RIGHTS RESERVED. No part of this publication may be reproduced or transmitted in any form whatsoever, electronic, or mechanical, including photocopying, recording, or by any informational storage or retrieval system without express written, dated and signed permission from the author.

Table of Contents

PYTHON FOR BEGINNERS 1
   A CRASH COURSE GUIDE TO LEARN PYTHON IN 1 WEEK..................1
INTRODUCTION 5
CHAPTER 1: PYTHON: A COMPREHENSIVE BACKGROUND 7
   ASSOCIATIONS THAT USE PYTHON..9
CHAPTER 2: HOW TO DOWNLOAD AND INSTALL PYTHON....12
   THE PYTHON ENVIRONMENTAL VARIABLES TO NOTE 22
CHAPTER 3: PYTHON GLOSSARY 24
CHAPTER 4: INTERACTING WITH PYTHON 26
   HOW TO OPEN A CONSOLE IN LINUX 27
   HOW TO OPEN A CONSOLE ON WINDOWS 27
   STARTING AND INTERACTING WITH PYTHON..............................28
CHAPTER 5: USING TURTLE FOR A SIMPLE DRAWING..........32
CHAPTER 6: VARIABLES 39
CHAPTER 7: LOOPS 42
CHAPTER 8: NATIVE PYTHON DATATYPES 47
   1: PYTHON STRINGS 47
   PYTHON STRING FORMATTING 54
   FORMATTING STRINGS USING THE FORMAT () METHOD 56
   THE COMMON PYTHON STRING METHODS 57
   2: PYTHON LISTS 58
   HOW TO REMOVE OR DELETE LIST ELEMENTS 65
   THE PYTHON LIST METHODS 68
   CREATING LISTS ELEGANTLY THROUGH LIST COMPREHENSION 69
CHAPTER 9: PYTHON DICTIONARIES 71
   HOW TO ACCESS ELEMENTS OF THE DICTIONARY 72
   MODIFYING DICTIONARIES 76

CHAPTER 10: BOOLEAN LOGIC AND CONDITIONAL STATEMENTS 81
    BOOLEAN OPERATORS FOR 'FLOW CONTROL' 89
    CONDITIONAL STATEMENTS: IF AND ELSE STATEMENTS 90
    NESTED IF STATEMENTS 98
CHAPTER 11: CONSTRUCTING 'WHILE' LOOPS IN PYTHON .104
CHAPTER 12: CONSTRUCTING 'FOR LOOPS' IN PYTHON PROGRAMMING 113
    USING RANGE () IN FOR LOOPS 114
    USING SEQUENTIAL DATA TYPES IN FOR LOOPS 117
    THE NESTED FOR LOOPS 120
CHAPTER 13: CONSTRUCTING CLASSES AND DEFINING OBJECTS 124
CONCLUSION 133
DID YOU ENJOY THIS BOOK? 134

Introduction

Thank you for downloading this book.

The book has lots of actionable information that will get you started on the journey to becoming a pro at python programming.

Without doubt, code is the language of the future. Think about it, with the ever growing dependency on computers, one of the most critical things that anyone in the world today can learn is how to speak to computers in a language that they understand and have them do anything that you want to do. This skill is one whose demand is increasing at an increasing rate the world over. Whether you want to learn code to pursue a career as a programmer, web developer or graphics designer or want to learn code to be able to develop your own web applications and other computerized systems, the one thing you need to do is to start somewhere. What better place to start than to learn python?

Python is a simple yet powerful programming language that can enable you to start thinking like a programmer right from the beginning. It is very readable and the stress many beginners face about memorizing arcane syntax typically presented by other programming languages will not affect you at all. Conversely, you will be able to concentrate on learning concepts and paradigms of programming.

This book shall introduce you to an easy way to learn Python in just 7 days and in this time, be able to complete your own projects! By reading the book and implementing what you learn herein, you will realize just why major institutions like NASA, Google, Mozilla, Yahoo, Dropbox, IBM, Facebook and many others prefer to use python in their core products, services and business processes. Let's begin.

Thanks again for downloading this book. I hope you enjoy it!

Before we can get to a point of learning the ins and outs of python programming, let's start by building an understanding of what python programming is and what it is about.

Chapter 1: Python: A Comprehensive Background

Before we discuss Python programming, especially what python is, let me briefly say something about programming.

Being a good programmer does not simply entail just knowing a vast array of programming languages or how to code fast programs. Instead, it is about:

- Comprehending a problem abstractly and being able to change it into code.

- Looking for new ways to tackle, for instance, a scientific problem and knowing the kind of tools to use.

- Being able to fix a program when it is not working

- Writing a program that is quick enough, not the quickest possible

- Writing a program that other people can understand in a short period of time

Take note of the last point: Reproducible research and open science is becoming the norm in some research fields. This means that other people will probably have to read your code, understand exactly what you are doing and be able to recreate the code so that they can run it themselves.

By the end of this book therefore, we will not only be interested in the correctness of your solutions; we shall also look at whether we can understand how the program you create solves the problem.

This is to mean that you should always write your codes assuming that someone else will read it.

Let's go back to understanding python, the programming language of the future.

What Is Python?

In its simplest terms, python is a general-purpose, multi-paradigm, and interpreted programming language that gives programmers the ability to use various styles of programming to create complex or simple programs, get results faster, and write code in a way that resembles human language (explanation below).

Python is the programming language often used to create algorithms for sorting and analyzing chunks of data that businesses and organizations from all over the world collect.

The explanation above brings about some very interesting points about python:

1. *It's a high level language*

Python is a high-level language. This means the code you type to build a program is more like a human language than the typical code created to control machines. This, for one, makes things a lot simpler for you, the programmer, and means that someone else is better placed to understand the code if he/she wants to use it him/herself. The human-like (high-level) code then goes through a software called an interpreter that converts it into machine code, a language that machines can understand.

2. *Its open source*

The software that lets us make programs in Python is open source. This means it is available in the public domain and anyone can freely use it. The greatest advantage of this software is the fact that you can modify it and create your own version to perform particular tasks. This is actually the main reason why many people have openly embraced the open source concept and the use of python is no exception.

Let's discuss this briefly:

Associations that Use Python

Many organizations currently use Python to complete major tasks. You will not always hear about their uses since organizations are somewhat reserved about sharing their systems' information, or 'trade secrets.' Nonetheless, Python is still there making a great difference in the way organizations function and many common systems and applications have settled for Python for their development. Some of them include YouTube, Google Search, BitTorrent, NASA, Eve Online, iRobot machines, Yahoo, Facebook, Maya and many others.

Look at the following commercial uses of Python:

Corel: Over the years, people have used products such as PaintShop Pro to grab screenshots, modify pictures, draw fresh images, and perform many other graphic oriented tasks. What is amazing about this popular product is that it heavily relies on Python scripting. This means to automate tasks in PaintShop Pro, you will need a degree of Python knowledge.

D-link: It can be quite problematic to upgrade firmware over a network connection, and this company (D-link) encountered a situation where every upgrade tended to tie up a machine, something they described as a weak utilization of resources. Additionally, a number of upgrades needed extra work since there were problems with the targeted device.

The use of python to build a multi-threaded application that allows for the movement of updates to the devices enables *one machine* to service *several devices,* and a new methodology supported by python decreases all the reboots to one, after the installation of that fresh firmware is done.

Moreover, this company opted for Python over other languages such as Java because Python offers an easy to use serial communication code.

ForecastWatch.com: Have you ever wondered if someone somewhere reviews your weatherman's performance? Well, if you have, look no further; this company does that. This company compares the thousands of weather forecasters produced every day against real climatological data in order to determine their accuracy. The produced report advances weather forecasts.

In this particular case, the software used in the comparisons is a pure Python program since it contains standard libraries that are important in the collection, parsing, and storage of data from online sources. Additionally, the enhanced, multithreaded nature and capability of python gives it the ability to gather the forecasts from about 5,000 sources per day. The code is also a lot smaller than what other languages like PHP or Java would need.

Many other companies, softwares, and programs use Python and to sum this up, I will say that scientists, business people, teachers, governments, and even religious organizations, just to mention the least use python programming in one form or the other.

You too should also get started with learning python if you don't want to be left behind. We'll start from the beginning i.e. downloading the program.

Chapter 2: How to Download and Install Python

We will now discuss how to download Python on your Widows OS and Mac OS X.

Windows OS

If you are a Windows operating system user, you have to note that the program does not arrive prepackaged with windows. As such, you have to install it and ensure you install the latest version.

A number of years ago, Python saw an update that caused a split in the program. This split has become a challenge to many newcomers. However, you should not worry because I will take you through the installation of both versions and talk about some of their important features.

The first thing you have to do is go to the python download page. You'll actually note this division and also see the repository ask whether you want the latest Python 3 (3.6.1) release or Python 2 (2.7.13).

The first question you may ask yourself is which version you should download. While newer means better, naturally, being a Python programmer means you have to make your choice based on your end goal. For instance, python 2.7 is ideal for hobby projects like games (coded in Python) in which you want to introduce new features. If you want to get some project whose extension ends in 'py' running, the best version you can use is Python 2.7.

On the other hand, if your focus is learning Python, you should highly consider installing both versions, something very possible and easy; it also requires minimal risk and very little setup trouble. This will also give you an opportunity to work with the latest Python version and be able to run older scripts of the language while testing for backwards compatibility for fresh projects. Please visit wiki.python.org to get more information about the differences between the two versions.

If you are completely sure you only need a specific version, you can go ahead and download one of them. Just note that you will see an 'x86–64' executable installer under the two versions' main entry.

- Python 3.6.1 - 2017-03-21
  - Download Windows x86 web-based installer
  - Download Windows x86 executable installer
  - Download Windows x86 embeddable zip file
  - Download Windows x86-64 web-based installer
  - **Download Windows x86-64 executable installer**
  - Download Windows x86-64 embeddable zip file
  - Download Windows help file

The installer will immediately install the suitable 64/32 bit version on your PC. You can also read here for more information about the differences between a 64-bit and 32-bit version.

How to Install Python 2

The installation of this particular version is simple and the installer even sets up the variable that specifies the set of directories where the program is located in your PC known as PATH variable. You only have to run the installer, select 'install for all users' then select 'next'.

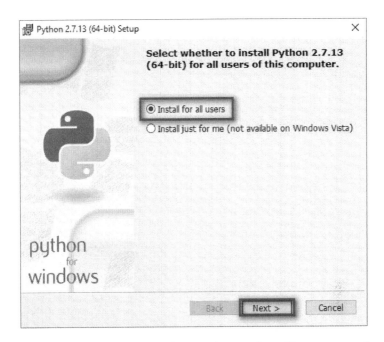

Leave the directory as 'Python27' on the directory selection screen and then click 'next.'

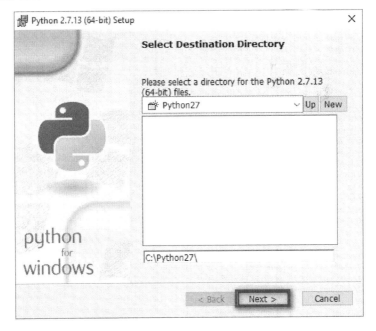

On the window, scroll down and select 'add python.exe to path,' and then click 'will be installed on local hard drive.' When finished, click next.

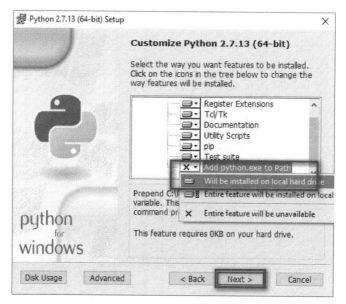

After the part above, you will realize you do not have to make any more decisions. Make sure to click through the wizard until you complete the installation. Once complete, you can now open up the command prompt in order to confirm the installation, and then type the following command: **python -V**

```
C:\Users\Jason>python -V
Python 2.7.13
```

Good job so far! If you wanted 2.7 to complete a pending project, you can stop here. The program is now installed on your device and its path variable set.

How to Install Python 3

Installation of the latest version of Python requires you complete this step, and as I mentioned earlier, you can install this version alongside python 2.7 without experiencing any problems. Go ahead and run the executable now. On the first screen, enable the option labeled 'add python 3.6 to PATH' and then click on 'install now'.

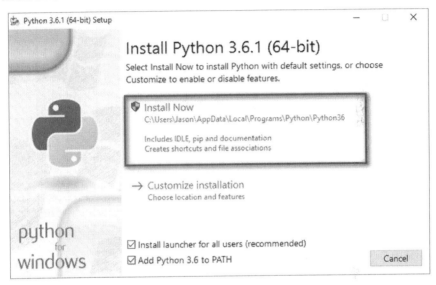

Once you do that, prepare yourself for an important decision. When you select the option 'disable path length limit', you will be taking out the limitation of the MAX_PATH variable. This does not change or distort anything. Actually, it allows the software to use lengthy path names. Currently, we have many programmers working in *nix systems such as Linux where the math name length is largely a negligible issue; therefore, you can turn this one on early enough to help you smooth over whichever issues related to path you may come across as you work in windows.

If you ask me, just go ahead and choose this option. If you are sure you do not want to restrict the path length limit, you can click close and complete the installation there.

In case you are just installing python 3, you can also use the command line trick-'python v' that we used above to confirm its proper installation and that the path variable is set.

How to Install Python on Mac OS X

Mac OS X typically comes with a pre-installed version of Python. For example, Mac OS X 10.8 comes with version 2.7. You only have to download and install the latest version alongside the one in the system to start enjoying the benefits of the most recent version of this program. Just go to python's official website to get the latest Python 3 version. Remember; you cannot depend on Python 3 to run python 2 scripts.

The Steps

When you use your Mac computer to access the Python download page, it will automatically sense that you are a Mac user and the options given henceforth will reflect that. The steps are simple and in a short while, you will download a file that carries the label '3.x.macosx10.6dmg' that will go directly to your browser's download folder.

When you open the file you downloaded, it will instantaneously mount on your desktop as a volume. Open the file and wait to see some window resembling the one below.

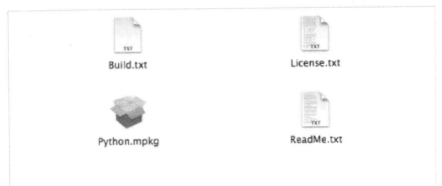

Before you proceed to install the program, take a few seconds to go through the 'readme.txt' file to learn much including that you may not just double click the installer labeled as

'Python.mpkg' since Apple does not sign it. The settings of your gatekeeper will however determine whether you can double click it or not, and option you will find under systems **preferences>privacy>general**. Ignore the window that appears after you are done, and proceed to the next step.

If your computer is using the gatekeeper default settings, proceed to right click on the installer package and click the option 'open with- installer.app' like this one:

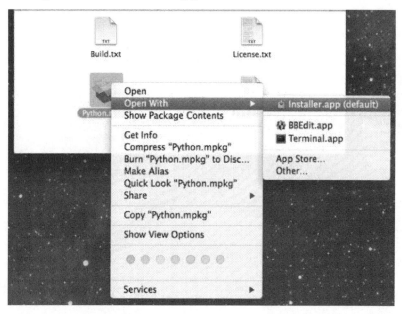

When you complete that, the system will take you through the standard process of installation and the steps are similar to the ones we covered in the Windows section.

Now that you have the program installed and running on your PC, let us go through two very important segments before we continue.

Setting up the PATH

Executable files and programs can be in multiple directories, so operating systems do give a search path listing the directories the operating system looks for executables. The path is kept in an environment variable, a named string that the operating system maintains. This variable has data open to the command shell and the other programs.

In Unix, the path variable is named as PATH (it's case-sensitive) while in Windows, it's referred to as Path (not case-sensitive). The installer in Mac OS manages the path details. If you want to invoke the Python interpreter from any specific directory, just add to your path the Python directory.

How to Set Up Path in Linux/Unix

To add the python directory to the path in Unix for a session, do the following:

1. Type the following in the csh shell: setenv PATH "$PATH:/usr/local/bin/python" and click enter.

2. Type the following in the Linux bash shell export ATH="$PATH:/usr/local/bin/python" then click enter.

3. Type the following in the ksh or sh shell: PATH="$PATH:/usr/local/bin/python" then click enter.

4. Remember that /usr/local/bin/python is the python directory path.

How to Set Up Path in Windows

To add the python directory to the path for a specific session in windows, do the following:

1. Type (at the command prompt) path %path%;C:\Python then click enter.
2. Remember that C:\Python refers to the python directory path.

The Python Environmental Variables to Note

Environmental variables in this regard are system-wide settings living outside python, which you can use to customize the interpreter's behavior every time you ran it on your computer.

It is important you know about them before we run our python program.

Python recognizes a number of environment variables contained in the following list:

PYTHONPATH

This one has a role that resembles PATH i.e. informing the Python interpreter the location where it should locate module files, which are usually imported into a given program. It usually contains the python source code as well as the Python source library directory.

PYTHONSTARTUP

It has the path of an initialization file that contains the Python source code. It executes each time you start the interpreter.

PYTHONCASEOK

If you are a Widows user, you will use this variable to ask Python (in an import statement) to search for the first case-insensitive match.

PYTHONHOME

This one is an alternative module search path typically embedded within the PYTHONPATH directory or the PYTHONSTARTUP directory to ease the process of switching module libraries.

By now, you must have come across a number of terms that you are unfamiliar with. To make it easy to understand these and the others that we will discuss as we go on in the book, let's briefly talk about a list of some of these terms.

Chapter 3: Python Glossary

Here is an explanation to some of the common terms that you will find in this book.

Augment: This refers to the supplementary information a computer uses to execute commands.

Class: A class is the template we use to build user defined objects.

Continue: Continue is a function we use to skip an existing block and turn back to the 'while' or 'for' statement

Conditional statement: A conditional statement is a statement containing an 'if/else' or 'if'.

Debugging: Debugging is the process of pursuing and purging errors in programming.

Def: Def is a function that mainly defines a method or function.

Iteration: Iteration refers to the process where instructions or structures are repeated sequentially a set number of times or until some condition is met.

Syntax: A syntax is a set of rules that outline how a python program should be written and interpreted.

Index: Index refers to the position in a well-ordered list assigned to the characters. For instance, each character holds an index beginning from 0 to the length of -1.

Variable: A variable is a reserved memory location a computer uses to store values.

Module: Refers to an object that contains attributes, arbitrarily named, that can reference and bind.

Console: Refers to the terminal where you can execute a command at a time.

Instantiate: The process of generating an instance. An instance is a specific realization of a template, a class of objects or a computer process for example.

Indentation: Refers to the placement of text away to the left or right in order to separate it from the adjacent text.

With that understanding, let's now get to the specifics of using python starting with interacting with python now that you have installed it on your computer.

Chapter 4: Interacting with Python

In this section, you will learn how to use and interact with Python in the number of available ways. Let us begin by talking about the python's interpreter, while using the console of your operating system.

A console is also referred to as a command prompt or a terminal. It is a textual method of interacting with your operating system just as the 'desktop' together with the 'mouse' is the graphical method used to interact with your PC.

How to Open a Console on Mac OS X

The standard console of OS X is a program known as "terminal". To open it, navigate to **applications**, then go to **utilities** and double click the **terminal** program.

You can also easily search for it in the search tool at the top right of the screen. The command line terminal is a tool you will use to interact with your computer.

A window having a command line prompt message that looks like the code below will open:

mycomputer:~ myusername$

How to Open a Console in Linux

The different distributions of Linux (such as mint, Fedora, Ubuntu) may have dissimilar console programs, typically known as terminals. The particular terminal you open and how you do so can depend on your distribution.

Let us take one example on Ubuntu; you will probably want to open Gnome Terminal—it presents a prompt similar to this:

myusername@mycomputer:~$

How to Open a Console on Windows

Window's console is also referred to as the command prompt, or cmd. You can easily get to it by using the key combination 'windows+R' (windows in this case refers to the windows log button), that opens the Run dialog. Just type cmd and press enter or just click okay.

You can also use the start menu to search for it. It should appear like this:

C:\Users\myusername>

The windows command prompt is not as potent as its counterparts on OS X and Linux are; therefore, you should consider directly starting the Python interpreter or use the IDLE program that comes with Python. All these are accessible from the Start menu.

Starting and Interacting With Python

The python program you have installed will act as something we usually refer to as an interpreter. The interpreter picks up text commands and runs them as you write them, which is quite handy for trying stuff out.

At your console, just type 'python' then press 'enter' and you should enter the interpreter.

As soon as Python opens, you will see some contextual information resembling the following:

```
Python 3.5.0 (default, Sep 20 2015, 11:28:25)
[GCC 5.2.0] on linux
Type "help", "copyright", "credits" or "license" for more information.
>>>
```

On the last line, we have the prompt >>> which indicates that you are currently in an interactive Python interpreter session known as 'python shell'. You should note that the python shell is different from the normal command prompt. At this point, you can try entering some code for Python to run. Try the following:

print("Hello world")

Now press enter and see the result. When you see the results, Python will take you back to the interactive prompt where you can enter another command:

```
>>> print("Hello world")
Hello world
>>> (1 + 4) * 2
10
```

'Help' is a very useful command since it enters a help functionality that lets you explore all the things python enables you to do, right from the interpreter. To close the help window, press 'q' and go back to the Python prompt.

You can also press Ctrl+Z to leave the interactive shell and return to the console or the system shell. On OS X or Linux, press Ctrl+D, or on the Windows button, enter.

Let us now try a simple exercise:

Earlier, I demonstrated entering a command to work out some math. You can now try out some math commands. Do you know any python operations? Tell it to give you the squared result of the sum of 239 and 588.

There are several ways you can get the answer:

```
>>> 239 + 588
827
>>> 827 * 827
683929
>>> (239 + 588) * (239 + 588)
683929
>>> (239 + 588) ** 2
683929
```

Running the Python Files

When you have a large python code to run, you will want to save it into a file. For example, you can modify it into little parts (fix a bug) and re-run the code without having to type the rest repeatedly. You can save your code to a file and pass the name of the file to the python program instead of typing the commands one by one. This executes the file's code instead of launching the interactive interpreter.

Let us try that:

Just create a file in the current directory (labelled 'hello.py') with your most preferred code editor and then enter the print command above. Next, save the file. On OS X or Linux, run 'touch hello.py' so that you create an empty file to edit. It is very easy to run this file with Python:

$ python hello.py

Ensure you are positioned at your system command prompt, which will either have $ or > at the end, not at python's (which contains >>> instead).

If you are using Windows, you can double click the file to run it.

When you press enter now, the file will execute and you will see the output. This time however, when python has completed the execution of all commands from the file, it exits back to the system command prompt as opposed to returning to the interactive shell.

At this point, we can now get started with the turtle project. Even so, you have to note the following:

NOTE: If you are getting weird errors about 'no such file or directory' or 'can't open file' instead of getting 'hello world', it means your command line is most probably not running from the directory in which you saved your file. You can go ahead and change your current command line's working directory with the 'cd' command—cd stands for 'change directory'. On windows, you might prefer something like:

> *cd Desktop\Python_Exercises*

If you are using OS X or Linux, you might prefer seeing something like:

$ *cd Desktop/Python_Exercises*

This changes to 'python_exercises' directory under the desktop folder (or somewhere like that). If unsure of the location directory you saved the file in, simply drag the directory to the command line window. Again, if you are not sure in which directory your shell is currently running, use pwd—it stands for 'print working directory'.

NOTE: When you begin playing around with turtle, avoid naming your file turtle.py. You can try using more apt names like 'rectangle.py' or 'square.py'. Otherwise, each time you refer to 'turtle', python will immediately pick up your file in place of the standard python 'turtle' module.

Before we learn a few more things about python and start handling the intermediate projects, you ought to be able to handle a few basic projects by now, including turtle. Let us cover that one first.

Chapter 5: Using Turtle for a Simple Drawing

Turtle is a python feature resembling a drawing board that lets you command some turtle to draw over it. In this feature, you can use functions such as turtle.left(...) and turtle.forward(...) which can move the turtle around.

Before we can use the turtle, we have to import it first. I recommend playing it in the interactive interpreter first since there is a bit more work needed to make it work from the files. To do so, go to your terminal and type the following:

Import turtle

➤

If you are using Mac OS and are not seeing anything, you can try issuing a command such as turtle.forward(0)' and check whether a new window opens behind your command line.

If you are using Ubuntu and receive the following error message 'no module named _tkinter', it means you have a missing package; just install it with 'sudo apt-get install python3-tk'

While it might be tempting to copy-paste some of the things written in this book into your terminal, I recommend that you type out each command because typing will get the syntax under your fingers (growing the muscle memory) and even assist avoid strange syntax-based errors.

turtle.forward(25)

→

turtle.left(30)

▼

The function 'turtle.forward(...)' will tell the turtle to move forward by a specific distance while 'turtle.left(...)' takes the number of degrees you desire to rotate to the left. We also have the 'turtle.right(...)' and 'turtle.backward(...)' too.

If you want to start afresh, simply type 'turtle.reset()' to clear the drawing your turtle created so far. Do not worry about this though; we will go into more on that shortly.

The standard turtle is a triangle; that is no fun. Instead of using the 'turtle.shape ()' command, let us try making it a turtle:

turtle.shape("turtle")

That is so much better.

If you placed the commands into a file, you should have noted the turtle window disappear after the turtle completed its movement. This is because python normally exits as soon as your turtle has completed moving. Since the turtle window is innate to python, it will go away as well. To prevent that, you can just put 'turtle.exitonclick()' beneath your file. The window remains open until you click on it.

import turtle

turtle.shape("turtle")

turtle.forward(25)

turtle.exitonclick()

In Python programming, text's horizontal indenting is important. We will learn all about it as we look at functions later on. For now though, let us consider that stray spaces or tabs can bring an unexpected error. You can even try adding one to see how badly python will react.

Draw A Square

NOTE: I do not expect you to always know the answer immediately. We all learn by trial and error. You need to experiment and see what python does when you feed it different commands, what gives beautiful results (even though sometimes quite unexpected), and what brings errors. It is also okay if you are not prepared to keep playing with something you have learned that generates fascinating results. Overall, do not think twice about trying (and probably failing) and learn something from it.

Let us try the following exercise:

Draw a square that looks like the one below:

For this square, perhaps you will require a right angle that is 90 degrees. This is how you can do this:

```
turtle.forward(50)
turtle.left(90)
turtle.forward(50)
turtle.left(90)
turtle.forward(50)
turtle.left(90)
turtle.forward(50)
turtle.left(90)
```

You can look at how the turtle begins and ends in the same place while facing the same direction before and after drawing the square. This is a very useful convention to follow, and it makes it easier to draw many shapes later on.

If you want to be creative, you can modify the shape using the 'turtle.color(...)' and 'turtle.width(...)' functions. How do you use these functions?

Before you use a function, you need to know its signature. For instance, what to put between the parentheses, and what those things even mean. To find that out, simply key in 'help(turtle.color)' into your Python Shell. If there is a large piece of text, Python will place the text into a pager, which will let you page up and down. To quit (and exit the pager), press the 'q' key.

NOTE: If you are getting an error such as this one: "NameError: name 'turtle' is not defined" when trying to view help, you ought to import names into python before you even refer to them. To do this, in a fresh interactive shell, you will want to import turtle before keying in help(turtle.color).

Alternatively, you can read more about functions on this page. Remember that if you actually misdrew something, you can simply tell turtle to erase the drawing board using the command 'turtle.reset()' or simply undo your most recent step using 'turtle.undo()'

As is detailed in help, you can easily modify the color using turtle.color(colorstring). Among many others, these include 'violet', 'green', and 'red.' For a more extensive list, see this manual.

Do you want to set an RGB value? Just make sure you run 'turtle.colormode (255) first.' After that, you could maybe run turtle.color(215, 100, 170) in order to set a pink color.

Draw a Rectangle

Let us now try drawing a rectangle.

Here is how you do it (the python code):

```
turtle.forward(100)
turtle.left(90)
turtle.forward(50)
turtle.left(90)
turtle.forward(100)
turtle.left(90)
turtle.forward(50)
turtle.left(90)
```

Also, note that when it comes to a triangle, and specifically, an equilateral triangle whose all sides are of equal length every corner has an angle of exactly 60 degrees.

Extra squares

We will now try drawing a tilted square, another one, and another one. It is up to you to decide how to experiment with the angles between the specific squares.

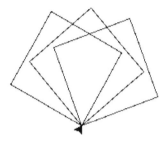

In the picture above, we have three turns of 20 degrees. You could try 30 or even 40-degree turns. For example:

```
turtle.left(20)

turtle.forward(50)
turtle.left(90)
turtle.forward(50)
turtle.left(90)
turtle.forward(50)
turtle.left(90)
turtle.forward(50)
turtle.left(90)

turtle.left(30)

turtle.forward(50)
turtle.left(90)
turtle.forward(50)
turtle.left(90)
turtle.forward(50)
turtle.left(90)
turtle.forward(50)
turtle.left(90)

turtle.left(40)

turtle.forward(50)
turtle.left(90)
turtle.forward(50)
turtle.left(90)
turtle.forward(50)
turtle.left(90)

turtle.forward(50)
turtle.left(90)
```

With that understanding of python, let's now move on to something even more eye opening about python i.e. variables.

Chapter 6: Variables

I'm sure you have noted that experimenting with angles requires changing three different numbers or places in the code each time. Can you imagine trying out all the sizes of squares or with rectangles? Would that not be tedious? Luckily, we have easier ways to do that than changing many numbers each time.

Using variables, we can achieve that sleekly. You will be able to tell python that whenever you refer to a variable, you actually mean something else. When you relate it to symbolic math, where you can write "let x be 5", this concept may be a bit more familiar. In this case, then x*2 is obviously 10.

In python syntax, this statement is explained as x = 5.

After the statement, if you happen to 'print (x)', it outputs the value -5. We can also use that for turtle as well.

turtle.forward(x)

Variables store all sorts of stuff, not just numbers. Another thing you may want to store regularly is a 'string'. A string just refers to a piece of text. These (Strings) usually have a starting as well as an ending double quote (") - we will delve into that, the other data types you can store, and what you can use these for later in the book.

Did you know that you can even refer to the turtle by name using a variable? Here is an explanation of this in play:

john = turtle

Now, each time you type 'john', python thinks you mean 'turtle'. However, you can keep using turtle as well:

```
john.forward(50)
john.left(90)
turtle.forward(50)
```

The Angle Variable

This is an exercise. If you create a variable known as tilt (you could assign it a number of degrees), how could you use that to make your experiment with the tilted squares program much faster?

This is the solution:

```
tilt = 20

turtle.left(tilt)

turtle.forward(50)
turtle.left(90)
turtle.forward(50)
turtle.left(90)
turtle.forward(50)
turtle.left(90)
turtle.forward(50)
turtle.left(90)

turtle.left(tilt)
```

- ..and so on!

NOTE: You could also apply this principle to the size of the squares

Santa Claus' House

Now draw a house

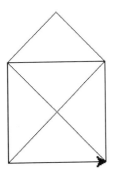

NOTE: You can calculate the diagonal line's length using the Pythagorean Theorem. That is actually a good value to store in as a variable. To determine the square root of a given number in Python, you ought to first import the specific math module then proceed to use this function: 'math.sqrt()'. We exponentiate a number with the ** operator (thus, squaring means **2).

import math

c = math.sqrt(a**2 + b**2)

Next, we will be discussing loops.

Chapter 7: Loops

You might have noticed something with our programs: there is often some repetition. Python has a strong concept it uses referred to as looping (the jargon is iteration). We will use it to cut all of our repetitive code. Meanwhile, try this simple example:

```
for name in "John", "Sam", "Jill":
   print("Hello " + name)
```

This can prove very helpful if you want to do something many times like when drawing the individual shape borderline but you only want to write the action just once. Look at this other version of loop:

```
for i in range(10):
   print(i)
```

You can see how we use 'i' to write only one line of code, but the values it takes are 10. You can consider the function 'range(n)' a shorthand for 0,1,2...,n-1. You can also use the help in the shell to know more about it; just type 'help(range)'. To exit the help again, use the 'q' key. You can also loop over all the elements you want:

```
total = 0
for i in 5, 7, 11, 13:
    print(i)
    total = total + i

print(total)
```

Write this out, run it, and check how it works.

As you can note above, the very lines of code indented are the same ones looped. In python, this is an important concept that allows it to know which lines should be used in the 'for loop' and the ones that come after, as part of the remaining section of the program. To indent your code, use four spaces (by hitting tab).

Sometimes, you may want to repeat some code a particular number of times but do not really care about the specific value of the 'i' variable; it can be good to replace it using _ instead. This shows that you do not care about its value, or do not even desire to use it. Look at this simple example:

```
for _ in range(10):
    print("Hello!")
```

I'm sure you are wondering about the variable 'i' and why we've used it all the time above. This variable stands for 'index' and is one of the most common variable names you will ever find in code. Even so, if you are looping over anything other than just numbers, you can even name it something better- for example:

```
for drink in list_of_beverages:
    print("Would you like a " + drink + "?")
```

You may notice that this is immediately clearer to read than if we had used 'i' instead of 'drink'.

The Dashed Line

Now draw a dashed line. You can use the turtle.penup () function to move the turtle without it drawing its movement. Use turtle.pendown () to instruct it to draw again.

```
for i in range(10):
    turtle.forward(15)
    turtle.penup()
    turtle.forward(5)
    turtle.pendown()
```

NOTE: you can make the dashes bigger as the line progresses.

If you are feeling lost, inspect 'i' at every loop's run.

```
for i in range(10):
    print(i)
    # write more code here
```

NOTE: You can make use of 'i' (or the index variable/loop variable) to get increasing step sizes.

Comments

In the above example, a comment is the line that begins with #. In this program, the computer ignores everything that goes a line after #. You can use comments to explain what your program does without altering the behavior for the computer. You can also easily use them to 'comment out' or disable some lines of code.

The comments can also be at the end of a line; look at this:

turtle.left(20)   # tilt our next square slightly

More squares

The squares we were drawing at the beginning had many repeated lines of code. You can try writing out a square drawing program in few lines by using loops.

Here is the python code for that:

```
for _ in range(4):
   turtle.forward(100)
   turtle.left(90)
```

You can try 'nesting' loops by placing one under (within) the other, with a drawing code that is inside the two. This is how it can appear:

```
for ...:
   for ...:
      # drawing code inside the inner loop goes here
      ...
   # you can put some code here to move
   # around after!
   ...
```

Now replace the…'s with a code of your own to see if you can come up with something different or interesting.

Loops aside, the next part will focus on native python datatypes.

Chapter 8: Native Python Datatypes

To have a better understanding of all we have discussed so far (and what is coming), I will introduce a new topic that will go into more detail about specific basic components of python. Besides that, we will also discuss a lot more to make sure you can handle larger projects with the program.

1: Python Strings

We are now going to discuss how to make, format, change, and delete python strings. You will also learn about the various string operations and functions.

A string is a sequence of characters. A character is a symbol; for instance, we have 26 characters in the English language. Nonetheless, computers do not deal with characters; they deal with binary or numbers. Even though we usually see characters on the screen, internally, they are stored and used as a mixture of zeros (0's) and ones (1's).

The conversion of characters to numbers is also known as encoding and some of the most popular encoding used are Unicode and ASCII (decoding is the reverse process). In Python, the sequence of Unicode character is known as string. At its introduction, encoding was meant to include every character in all languages and bring some uniformity in encoding. To learn more about encoding, read here. That aside, let's discuss how to make strings.

How to Make Strings

You can create strings by enclosing characters within single quotes or double quotes. You can even use triple quotes in Python but generally to represent docstrings and multiline strings.

```
# all of the following are equivalent
my_string = 'Hello'
print(my_string)


my_string = "Hello"
print(my_string)


my_string = '''Hello'''
print(my_string)


# triple quotes string can extend multiple lines
my_string = """Hello, welcome to
    the world of Python"""
print(my_string)
```

Accessing Characters in a String

You can access particular characters using indexing and an array of characters through slicing. Index starts from 0 and when you try to access a character out of index range, it causes an 'IndexError'. This index must be an integer. We cannot use float or other forms as it will result into TypeError.

In python, negative indexing is used for its sequences.

The -1 index refers to the final item, the second last item is represented by -2, and so on. When you use the slicing operator (colon), you can be able to access a range of items.

str = 'programiz'

print('str = ', str)

#first character

print('str[0] = ', str[0])

#last character

print('str[-1] = ', str[-1])

#slicing 2nd to 5th character

print('str[1:5] = ', str[1:5])

#slicing 6th to 2nd last character

print('str[5:-2] = ', str[5:-2])

If you try accessing index outside the range, or use a decimal number, you get errors.

```
# index must be in range
>>> my_string[15]
...
IndexError: string index out of range

# index must be an integer
>>> my_string[1.5]
...
TypeError: string indices must be integers
```

You can best visualize slicing by considering the index to remain between the elements as illustrated below. If you want to access a range, you will require the index that will help with slicing the portion from the string.

P	R	O	G	R	A	M	I	Z	
0	1	2	3	4	5	6	7	8	9
-9	-8	-7	-6	-5	-4	-3	-2	-1	

Changing and Deleting a String

Strings are largely immutable; this means you cannot change the string elements once you have assigned it. You can simply reassign various strings to the same name.

```
>>> my_string = 'programiz'
>>> my_string[5] = 'a'
...
TypeError: 'str' object does not support item assignment
>>> my_string = 'Python'
>>> my_string
'Python'
```

You cannot remove or delete characters from a string unless you are deleting the entire string itself; this is possible by using the 'del' keyword.

```
>>> del my_string[1]
...
TypeError: 'str' object doesn't support item deletion
>>> del my_string
>>> my_string
...
NameError: name 'my_string' is not defined
```

Next, we will discuss string operators.

The String Operators

We have numerous operations you can perform with string; this makes it one of the most used python datatypes. For instance, you can do the following:

1-Concatenation of Multiple Strings

A concatenation is the joining of two or more strings into one string. In Python, we use the + operator to achieve this. When you write two string literals together, it also concatenates them. We use the * operator to repeat the strings a particular number of times.

```
str1 = 'Hello'
str2 ='World!'


# using +
print('str1 + str2 = ', str1 + str2)


# using *
print('str1 * 3 =', str1 * 3)
```

When you write two string literals, this also concatenates them just like the + operator. You can also use parentheses if you want to proceed to concatenate strings in separate lines.

```
>>> # two string literals together
>>> 'Hello "World!"
'Hello World!"

>>> # using parentheses
>>> s = ('Hello '
...      'World')
>>> s
'Hello World'
```

2-Iterating Through String

You can also use 'for loop' to iterate through a string. For instance, you can count the sum of '1' in a string.

```
count = 0
for letter in 'Hello World':
    if(letter == 'l'):
        count += 1
print(count,'letters found')
```

3-The String Membership Test

You can test if a substring occurs inside a string or not by using the keyword 'in'. Here is an example:

```
>>> 'a' in 'program'
True
>>> 'at' not in 'battle'
False
```

4- The Built-In Functions to Use with Python

Many built-in functions that work with sequence also work with strings. Some of the most popular functions are len () and enumerate (). The enumerate () function takes back an enumerate object. It has the index and value of every item in the string as pairs, something that can be very useful for iteration.

Likewise, len () takes back the number of characters (length) of the string.

```python
str = 'cold'

# enumerate()
list_enumerate = list(enumerate(str))
print('list(enumerate(str) = ', list_enumerate)

#character count
print('len(str) = ', len(str))
```

Python String Formatting

Here is how python strong formatting works:

Escape Sequence

Say you want to print some text such as 'he said, what's there', you cannot use single or double quotes. Since 'the' text itself has both double and single quotes, it will definitely result into 'SyntaxError'.

```
>>> print("He said, "What's there?"")
...
SyntaxError: invalid syntax
>>> print('He said, "What's there?"')
...
SyntaxError: invalid syntax
```

To dodge this problem, you can simply use triple quotes, or instead, use escape sequences. An escape sequence begins with a backlash and it is interpreted differently. If you use single quotes to signify a string, all the single quotes within the text have to be escaped. This is similar to the case with double quotes. This is how you can do it to represent the text above.

```python
# using triple quotes
print('''He said, "What's there?"''')


# escaping single quotes
print('He said, "What\'s there?"')


# escaping double quotes
print("He said, \"What's there?\"")
```

The following is a list of the python-supported escape sequence:

Escape Sequence	Description
\newline	Backslash and newline ignored
\\	Backslash
\'	Single quote
\"	Double quote
\a	ASCII Bell
\b	ASCII Backspace
\f	ASCII Formfeed
\n	ASCII Linefeed
\r	ASCII Carriage Return
\t	ASCII Horizontal Tab
\v	ASCII Vertical Tab
\ooo	Character with octal value ooo
\xHH	Character with hexadecimal value HH

Look at the following examples:

```
>>> print("C:\\Python32\\Lib")
C:\Python32\Lib

>>> print("This is printed\nin two lines")
This is printed
in two lines

>>> print("This is \x48\x45\x58 representation")
This is HEX representation
```

Ignoring the Escape Sequence with Raw String

As you will soon realize, there are times when you will just wish to ignore the escape sequences within a string. To do this, you can place R or r before the string. This implies that this is actually a raw string and all escape sequences within are ignored.

```
>>> print("This is \x61 \ngood example")
This is a
good example
>>> print(r"This is \x61 \ngood example")
This is \x61 \ngood example
```

Formatting Strings Using the Format () Method

The format () method that is available with the string object is quite versatile and strong in string formatting. Format strings has curly braces {} as replacement fields or placeholders that gets replaced. You can use keyword arguments or positional arguments (explained later) to specify the order.

This 'format ()' method can have discretionary format specifications, and the colon separates them from the field name. For instance, we can right-justify, left-justify or center ^ a string in the set space. We also format integers as hexadecimal, binary and so on, and floats are presentable in the exponent format or rounded.

You can use numerous formatting methods. Visit Pyformat.info for all the available string formatting with the format () method.

```
>>> # formatting integers
>>> "Binary representation of {0} is {0:b}".format(12)
'Binary representation of 12 is 1100'

>>> # formatting floats
>>> "Exponent representation: {0:e}".format(1566.345)
'Exponent representation: 1.566345e+03'

>>> # round off
>>> "One third is: {0:.3f}".format(1/3)
'One third is: 0.333'

>>> # string alignment
>>> "|{:<10}|{:^10}|{:>10}|".format('butter','bread','ham')
'|butter    |   bread  |       ham|'
```

The Old-Style Formatting

You can actually format strings such as the old sprintf () that is popularly used in the C programming language. To achieve this, we use the % operator as shown below:

```
>>> x = 12.3456789
>>> print("The value of x is %3.2f %x)
The value of x is 12.35
>>> print("The value of x is %3.4f %x)
The value of x is 12.3457
```

The Common Python String Methods

There are various methods available with the string object; the format () method we talked about is one of them. Some of the most popular methods used include upper (), lower (), replace (), join (), split () etc. You can find a complete list of all the Python's built in methods to work with strings here.

```
>>> "PrOgRaMiZ".lower()
'programiz'
>>> "PrOgRaMiZ".upper()
'PROGRAMIZ'
>>> "This will split all words into a list".split()
['This', 'will', 'split', 'all', 'words', 'into', 'a', 'list']
>>> ' '.join(['This', 'will', 'join', 'all', 'words', 'into', 'a', 'string'])
'This will join all words into a string'
>>> 'Happy New Year'.find('ew')
7
>>> 'Happy New Year'.replace('Happy','Brilliant')
'Brilliant New Year'
```

From sequences of characters, let us now try looking at other data structures—all the different collections you have on your computer—the assortment of your files, browser bookmarks, your song playlists, emails, the video collections you can access on a streaming service and many more.

Let us talk about lists.

2: Python Lists

In this section, you will learn all about lists, how they are made, the process of adding or removing elements from them, and so forth.

Python offers many compound datatypes usually known as sequences. Apart from being one of the most frequently used datatype in Python, 'lists' is very versatile.

There are a number of ways to describe lists:

The usual description of a list type is "the container that holds other objects in a particular order." It executes the sequence protocol and lets you add or take out objects from the sequence.

Another description of the list is as a data structure that is a changeable, or mutable, organized sequence of elements. Every element or value within a list is referred to as an item. Just as strings are known as characters in between quotes, lists are characterized by values in between square brackets [ ].

When you want to work with many related values in Python, lists are very handy to use. They will allow you to keep data that belongs together together, condense code, and perform similar methods and operations on different values all at once.

When you are thinking about lists in python, and other data structures that are essentially types of collections, you can try considering all the different collections you have on your computer, the assortment of your files, browser bookmarks, your song playlists, emails, the video collections you can access on a streaming service, and many more.

Creating a List

In Python, you can create a list by placing the elements (or all the items) within a square bracket [], and separating with commas. It can contain any number of items that may be of different types (float, integer, string, etc.).

```
# empty list
my_list = []

# list of integers
my_list = [1, 2, 3]

# list with mixed datatypes
my_list = [1, "Hello", 3.4]
```

Additionally, a list can contain another list as an item in what we call a nested list.

```
# nested list
my_list = ["mouse", [8, 4, 6], ['a']]
```

Accessing Elements from a List

To access list elements, you can use a number of ways:

List index

You can use the index operator [] to get to an item in a list. Index begins from 0, and thus, a list of five items will contain an index from 0 to 4. When you try to access an element other than this, it raises an IndexError. The index should be an integer and we cannot use float or other types because the results will be a TypeError.

Nested list are accessed via nested indexing

```
my_list = ['p','r','o','b','e']
# Output: p
print(my_list[0])


# Output: o
print(my_list[2])


# Output: e
print(my_list[4])


# Error! Only integer can be used for indexing
# my_list[4.0]


# Nested List
n_list = ["Happy", [2,0,1,5]]


# Nested indexing


# Output: a
print(n_list[0][1])


# Output: 5
print(n_list[1][3])
```

Negative Indexing

Python programming supports negative indexing within its different sequences. As I mentioned while discussing string, the -1 index is used in reference to the final item, -2 is used in reference to second last one etc.

```python
my_list = ['p','r','o','b','e']
```

```python
# Output: e
print(my_list[-1])
```

```python
# Output: p
print(my_list[-5])
```

Slicing Lists in Python

We can access numerous items in a list with the colon/slicing operator.

```python
my_list = ['p','r','o','g','r','a','m','i','z']
# elements 3rd to 5th
print(my_list[2:5])
```

```python
# elements beginning to 4th
print(my_list[:-5])
```

```python
# elements 6th to end
print(my_list[5:])
```

```python
# elements beginning to end
print(my_list[:])
```

You can best visualize slicing by considering the index supposed to go between the elements as displayed below. Thus, if you want to access a range, you will need two indexes that will slice that particular portion from the list.

```
| P | R | O | G | R | A | M | I | Z |
  0   1   2   3   4   5   6   7   8   9
 -9  -8  -7  -6  -5  -4  -3  -2  -1
```

Changing or adding elements to a list

I mentioned in the beginning that lists are mutable; this means unlike tuple or strings, their elements can be changed. When changing an item or multiple items, we can use the assignment operator = to do that.

\# mistake values

odd = [2, 4, 6, 8]

\# change the 1st item

odd[0] = 1

\# Output: [1, 4, 6, 8]

print(odd)

\# change 2nd to 4th items

odd[1:4] = [3, 5, 7]

\# Output: [1, 3, 5, 7]

print(odd)

You can also use **append ()** method to add one item to a list or use the **extend ()** method to add multiple items.

odd = [1, 3, 5]

odd.append(7)

```
# Output: [1, 3, 5, 7]
print(odd)
```

odd.extend([9, 11, 13])

```
# Output: [1, 3, 5, 7, 9, 11, 13]
print(odd)
```

You can also combine two lists using the + operator. This is known as concatenation. The * operator on the other hand repeats a list for the set number of times.

odd = [1, 3, 5]

```
# Output: [1, 3, 5, 9, 7, 5]
print(odd + [9, 7, 5])

#Output: ["re", "re", "re"]
print(["re"] * 3)
```

Furthermore, you can insert a single item at a desired place by using the **insert ()** method or just insert many items by squeezing it into an empty list slice.

```
odd = [1, 9]
odd.insert(1,3)


# Output: [1, 3, 9]
print(odd)


odd[2:2] = [5, 7]


# Output: [1, 3, 5, 7, 9]
print(odd)
```

How to Remove or Delete List Elements

You can delete a single or multiple items from a list by using the 'del' keyword. It can actually delete the list completely.

```python
my_list = ['p','r','o','b','l','e','m']

# delete one item
del my_list[2]

# Output: ['p', 'r', 'b', 'l', 'e', 'm']
print(my_list)

# delete multiple items
del my_list[1:5]

# Output: ['p', 'm']
print(my_list)

# delete entire list
del my_list

# Error: List not defined
print(my_list)
```

You can also use **remove ()** to remove the given item or **pop ()** method to get rid of an item at the given index.

The pop () method removes and returns the final item if index is not given. This in particular helps us implement lists just like stacks (first-in, last-out structure). To empty a list, you can also use the **clear ()** method.

my_list = ['p','r','o','b','l','e','m']

my_list.remove('p')

# Output: ['r', 'o', 'b', 'l', 'e', 'm']
print(my_list)

# Output: 'o'
print(my_list.pop(1))

# Output: ['r', 'b', 'l', 'e', 'm']
print(my_list)

# Output: 'm'
print(my_list.pop())

# Output: ['r', 'b', 'l', 'e']
print(my_list)

my_list.clear()

# Output: []
print(my_list)

Lastly, you can also delete list items by allocating an empty list to a slice of elements.

```
>>> my_list = ['p','r','o','b','l','e','m']
>>> my_list[2:3] = []
>>> my_list
['p', 'r', 'b', 'l', 'e', 'm']
>>> my_list[2:5] = []
>>> my_list
['p', 'r', 'm']
```

The Python List Methods

In the following table are the methods available with the Python programming list object. You will access them as list.method (). Note that we have already used some of these methods above.

The list of python Methods
append() – use it to add some element to the end of the list
extend() – use it to add all list elements to the another list
insert() – use it to insert an item at the set index
remove() – use it to remove an item from your list
pop() – use it to remove and return some element at the given index
clear() – use it to removes all your items from the list
index() – use it to return the first matched item's index
count() – use it to returns the count of number of items that have passed as an argument
sort() – use it to sort list items in an ascending order
reverse() – use it to reverse the list's order of items

Look at some examples of python list methods:

my_list = [3, 8, 1, 6, 0, 8, 4]

# Output 1
print(my_list.index(8))

# Output 2
print(my_list.count(8))

my_list.sort()

# Output [0, 1, 3, 4, 6, 8, 8]
print(my_list)

my_list.reverse()

# Output [8, 8, 6, 4, 3, 1, 0]
print(my_list)

Creating Lists Elegantly Through List Comprehension

One very elegant and precise way of creating new lists from existing python lists is though list comprehension.

List comprehension comprises an expression that is followed by the 'for statement' within square brackets. The following is an example to make a list with every item assuming an increasing power of 2.

```
pow2 = [2 ** x for x in range(10)]


# Output: [1, 2, 4, 8, 16, 32, 64, 128, 256, 512]

print(pow2)
```

This code is equivalent to

```
pow2 = []
for x in range(10):
    pow2.append(2 ** x)
```

Optionally, the list comprehension can have more 'for' or 'if' statements. An optional 'if' statement will filter out items for the fresh list. Examine these examples:

```
>>> pow2 = [2 ** x for x in range(10) if x > 5]
>>> pow2
[64, 128, 256, 512]
>>> odd = [x for x in range(20) if x % 2 == 1]
>>> odd
[1, 3, 5, 7, 9, 11, 13, 15, 17, 19]
>>> [x+y for x in ['Python ','C '] for y in ['Language','Programming']]
['Python Language', 'Python Programming', 'C Language', 'C Programming']
```

Well, 'lists' just contains values. At this point, I think we should look at something that associates every value with a key: dictionaries. It is also a good time to let the order rest for a while!

Chapter 9: Python Dictionaries

In python, dictionaries refer to the unordered collection of items. We also refer to them as the built-in 'mapping' type of python. Dictionaries are helpful because they map keys to values and these pairs (key-value) offer a convenient way to store python's data. Thus, while other compound types of data contain only a single value as an element, the dictionary contains a key: *value pair*.

We optimize dictionaries to retrieve values when we know the key.

Dictionaries are made with curly braces on both sides since they have to hold related data like the information contained in an ID or a user profile. A dictionary simply looks like this:

sammy = {'username': 'sammy-shark', 'online': True, 'followers': 987}

Apart from the curly braces, the dictionary also contains colons (:) throughout.

The words to the left of the colons represent keys. Keys can comprise any immutable data type and in the dictionary above, the keys include the following:

- 'username'
- 'followers'
- 'online'

NOTE: Every key in the example above is a string value.

The words to the right of the colons represent the values. Values can consist of any data type. In the dictionary above, the values include the following:

- 'sammy-shark'
- True
- 987

Each one of these values is either an integer or Boolean (see next section).

Let us now print the 'sammy' dictionary:

```
print(sammy)
Output
{'username': 'sammy-shark', 'followers': 987, 'online': True}
```

When you look at the output, you may notice that the key-value pairs' order has changed. This is because of the 'unordered' nature of the dictionary data type. Therefore, unlike tuples and lists, dictionaries do not maintain order or the ability to be indexed. Any time you print a dictionary, the order is usually random. However, the key-value pairs usually remain intact all the time thus enabling people to access data according to their relational meaning.

How to Access Elements of the Dictionary

Due to the dictionary data structure's nature of being unordered, you cannot call its values by an index number as you can with tuples and lists. Nonetheless, you can reference the related keys to call its values.

-Accessing Data Items Using Keys

In your python program, dictionaries are very important because they provide key-value pairs to store data. If you want to seclude the username 'Sammy', you can do so by calling Sammy['username']. Let us try printing that out:

```
print(sammy['username'])
Output
sammy-shark
```

Like databases, dictionaries are dissimilar to lists in that as opposed to calling an integer in order to get a specific index value, you assign a value to a key and have the option of calling that key to get its related value.

When you invoke the 'username' key, you get the specific value of that particular key i.e. 'sammy-shark'. The values that linger within Sammy dictionary could actually be called using that same format:

```
sammy['followers']
# Returns 987

sammy['online']
# Returns True
```

When you take advantage of the key-value pairs of dictionaries, you can reference keys to retrieve values.

-Accessing Elements with Functions

Apart from accessing values using keys, you can also work with a number of built-in functions:

- ✓ dict.keys() isolates keys
- ✓ dict.values() isolates values
- ✓ dict.items() returns items in a list format of (key, value) tuple pairs

To return the keys, you would use the function 'dict.keys()'. In the example, that would take the variable name and thus be 'sammy.keys()'. We'll pass that to a 'print()' function and observe the output:

```
print(sammy.keys())
Output
dict_keys(['followers', 'username', 'online'])
```

We get output that puts the keys in an iterable view object of dict_keys class. The keys are thus printed inside a list format.

You can use this function to query across dictionaries. For instance, you can see the common keys shared between the dictionary data structures:

```
sammy = {'username': 'sammy-shark', 'online': True, 'followers': 987}
jesse = {'username': 'JOctopus', 'online': False, 'points': 723}

for common_key in sammy.keys() & jesse.keys():
    print(sammy[common_key], jesse[common_key])
```

The 'sammy' dictionary and the 'jesse' dictionary are both a user profile dictionary. Their profiles contain different keys. Nonetheless, since Sammy has a social profile carrying associated followers and Jesse carries a gaming profile that has associated points, both keys they have in common are 'online' status and 'username' that you can get when you run this little program:

Output
sammy-shark JOctopus
True False

You could definitely improve on the program to ensure the output is more readable to the user; however, this illustrates that you can use the 'dict.keys()' to check across different dictionaries to be able to see what they share in common or not. You will find this particularly helpful with large dictionaries.

You can also use the 'dict.values()' function in a similar way to query the values in the 'sammy' dictionary, which you would construct as 'sammy.values'. Let us try printing them out:

sammy = {'username': 'sammy-shark', 'online': True, 'followers': 987}

print(sammy.values())
Output
dict_values([True, 'sammy-shark', 987])

Both the 'values()' and 'methods()' return unsorted keys lists and values that exist within the 'sammy' dictionary with all the view objects of 'dict_keys' as well as 'dict_values' in that specific order. If you are interested in all the dictionary items, you can simply access them with the 'items()' function.

print(sammy.items())
Output
dict_items([('online', True), ('username', 'sammy-shark'), ('followers', 987)])

This 'lists' returned format usually has (key, value) tuple pairs containing dict_items view object.

By using a 'for loop', you can actually iterate the returned format list. For instance, you can print every one key and value of a particular dictionary and then make it more readable to any user by adding a string:

```
for key, value in sammy.items():
    print(key, 'is the key for the value', value)
Output
online is the key for the value True
followers is the key for the value 987
username is the key for the value sammy-shark
```

If you check above, the 'for loop' actually iterated over the items in the 'sammy' dictionary and then printed out the keys as well as values line after line, with information to make it simpler to understand by users.

You can utilize the in-built functions to access values, items and keys from the data structures of the dictionary.

Modifying Dictionaries

Dictionaries, as I mentioned earlier, are mutable structures of data that you can modify. In this regard, we will now look at adding and deleting elements in a dictionary.

Adding and Altering Elements

You could add key-value pairs without using a function by using the syntax 'dict[key] = value'. Let us engage in a practical look at how this works by adding a key-value pair to 'usernames' dictionary.

```
usernames = {'Sammy': 'sammy-shark', 'Jamie': 'mantisshrimp54'}

usernames['Drew'] = 'squidly'

print(usernames)
Output
{'Drew': 'squidly', 'Sammy': 'sammy-shark', 'Jamie': 'mantisshrimp54'}
```

As you can see, the dictionary has been updated with the key-value pair 'drew': 'squidly'. This pair can go anywhere in the dictionary since dictionaries are typically unordered. If you can use the dictionary 'usernames' later in your program file, it will contain the additional key-value pair.

Additionally, you can use this syntax to modify the value assigned to a key. In this case, you will reference a key that exists and pass to it a different value.

We will consider a dictionary 'drew' which is one of the users on a given network. Let us say that today, this user had a bump in followers. We have to update this integer passed to the key (followers). We will use the function 'print()' to confirm that the dictionary was indeed modified.

```
drew = {'username': 'squidly', 'online': True, 'followers': 305}

drew['followers'] = 342

print(drew)
Output
{'username': 'squidly', 'followers': 342, 'online': True}
```

In the output, we can see that the number of followers jumped to 342, from the integer value of 305.

We can utilize this method to add key-value pairs to dictionaries with the user-input. Let us now try writing a program (usernames.py), which will run on the command line, as well as allow input from the user to add in more names and linked usernames:

usernames.py

```python
# Define original dictionary
usernames = {'Sammy': 'sammy-shark', 'Jamie': 'mantisshrimp54'}

# Set up while loop to iterate
while True:

    # Request user to enter a name
    print('Enter a name:')

    # Assign to name variable
    name = input()

    # Check whether name is in the dictionary and print feedback
    if name in usernames:
        print(usernames[name] + ' is the username of ' + name)

    # If the name is not in the dictionary...
    else:

        # Provide feedback
        print('I don\'t have ' + name + '\'s username, what is it?')

        # Take in a new username for the associated name
        username = input()

        # Assign username value to name key
        usernames[name] = username

        # Print feedback that the data was updated
        print('Data updated.')
```

Let us now use python usernames.py to execute the program:

When you run it, you get this output:

Output
Enter a name:
Sammy
sammy-shark is the username of Sammy
Enter a name:
Jesse
I don't have Jesse's username, what is it?
JOctopus
Data updated.
Enter a name:

After testing the program, press Ctrl+C to escape it, and perhaps set up a trigger that enables you to close/quit the program—like typing the letter 'q'—along with a conditional statement to advance your code. You can interactively modify dictionaries using this approach. With this program in particular, when you press Ctrl+C to exit, you lose all the data unless you find a way to manage reading and writing files.

You can also use the 'dict.update()' function to add and modify dictionaries. This will vary from the 'append()' function available in lists.

We will now add the 'followers' key in the dictionary below and then give it an integer value in 'jesse.update()'. Let's 'print()' the updated dictionary to follow that.

jesse = {'username': 'JOctopus', 'online': False, 'points': 723}

jesse.update({'followers': 481})

print(jesse)
Output
{'followers': 481, 'username': 'JOctopus', 'points': 723, 'online': False}

You can see from the output that we added the key-value pair 'followers': 481 to the dictionary.

You can as well change the current key-value pair by using the 'dict.update()' method by simply replacing a specific value for a particular key.

Let us now change sammy's online status in the 'sammy' dictionary from true to false:

sammy = {'username': 'sammy-shark', 'online': True, 'followers': 987}

sammy.update({'online': False})

print(sammy)
Output
{'username': 'sammy-shark', 'followers': 987, 'online': False}

The 'Sammy.update({'online':False})' line reference the 'online' key that exists and modifies its Boolean (see next section) value from True to False. When you call to print() the dictionary, you see the update occur in the output.

You can either use the function 'dict.update()' or syntax 'dict[key]=value' to add items to dictionaries, or modify values.

Now, for the easiest part of dictionaries>> Deleting Elements; please go over this <u>information</u> to know how to do that with python.

At this point, you have just covered most of the basic programming topics. It is about time we took all this up a notch. Let us talk about something interesting; what do you think about designing a circuit that gives you an output only when particular combinations are existent? Well, this (and more) is what Boolean logic helps you do.

Chapter 10: Boolean Logic and Conditional Statements

The Boolean type of data can be one or two values, either false or true. In programming, we use Booleans to compare and control the flow of the program.

Booleans symbolize the true values linked with logic in mathematics, which forms the foundation of algorithms in computer science. Booleans are named after George Boole, the mathematician who lived over two centuries ago. Thus, the word Boolean should begin with a capitalized B. The False and True values also have to begin with capital F and T respectively because in Python, they are very special values.

Before we get deeper into the topic, we will start by going over the basics of Boolean logic, including how they work, the Boolean comparisons, truth tables and logical operators.

Comparison Operators

In programming, we use comparison operators to make comparisons of values and evaluate down to one Boolean value of either False or True.

The following is a table showing Boolean comparison operators.

Operator	What it means
==	Equal to
!=	Not equal to
<	Less than
>	Greater than
<=	Less than or equal to
>=	Greater than or equal to

To understand how the operators work in python, you have to assign two integers to two variables:

```
x = 5
y = 8
```

In the example above, x has a value of 5, and is less than y which takes the value of 8. Using the two variables and their connected values, we will go through the operators from the table above. In the program, instruct python to print out 'if' each comparison operator evaluates to true or false. We will have python print a string to show us what it is evaluating so we and other people can understand this output better.

```
x = 5
y = 8

print("x == y:", x == y)
print("x != y:", x != y)
print("x < y:", x < y)
print("x > y:", x > y)
print("x <= y:", x <= y)
print("x >= y:", x >= y)
```

Output
x == y: False
x != y: True
x < y: True
x > y: False
x <= y: True
x >= y: False

Python, following mathematical logic, has evaluated the following in each of the above expressions:

- Is 5 (x) equal to 8 (y)? **False**
- Is 5 not equal to 8? **True**
- Is 5 less than 8? **True**
- Is 5 greater than 8? **False**
- Is 5 less than or equal to 8? **True**
- Is 5 not less than or equal to 8? **False**

Even though we used integers here, we can also utilize float values to substitute them.

Boolean operators are also usable with strings; they are, however, case-sensitive unless you use an extra string method. Let us look at how strings are practically compared:

Sammy = "Sammy"
sammy = "sammy"

print("Sammy == sammy: ", Sammy == sammy)
Output
Sammy == sammy: False

The string above denoted by 'Sammy' is not equivalent to the 'sammy' string. This is because they are not the same: one begins with the upper case S while the other one begins with a lower-case s. However, if you added another variable assigned to the value of 'Sammy', they will then evaluate to equal.

```
Sammy = "Sammy"
sammy = "sammy"
also_Sammy = "Sammy"

print("Sammy == sammy: ", Sammy == sammy)
print("Sammy == also_Sammy", Sammy == also_Sammy)
Output
Sammy == sammy: False
Sammy == also_Sammy: True
```

You can also use the other comparison operators that include > and < to compare two strings. The program compares the strings using the ASCII values of the characters lexicographically.

You can also evaluate Boolean values with the comparison operators:

```
t = True
f = False

print("t != f: ", t != f)
Output
t != f: True
```

The above code block evaluates that True is not equal to False. Notice the dissimilarity between operators = and ==.

```
x = y   # Sets x equal to y
x == y  # Evaluates whether x is equal to y
```

This one '=' is the assignment operator that sets one value equivalent to another. The other one '==' is the comparison operator that evaluates whether both values are equal.

The Logical Operators

We generally use three logical operators to compare values. These operators evaluate expressions down to the Boolean values and return either False or True. These operators are 'or', 'not' and 'and'. Look at the table below to see their definition:

Operator	What it means	What it looks like
and	True if both are true	x and y
or	True if at least one is true	x or y
not	True only if false	not x

We typically use logical operators to evaluate whether expressions are true or not true. For instance, we can use them to determine if the student has registered for the course and that a grade is passing. If the two cases are true, the student will thus be given a grade in the system. Similarly, we can use it to determine whether someone is a valid active online shop customer based on whether he has a store credit or has purchased goods within the past six months.

To have a better understanding of how logical operators work, we will evaluate some three expressions:

```
print((9 > 7) and (2 < 4))  # Both original expressions are True
print((8 == 8) or (6 != 6)) # One original expression is True
print(not(3 <= 1))          # The original expression is False
Output
True
True
True
```

In the first scenario with 'print((9>7)' and '(2<4))', 9>7 and 2<4 required evaluation to True because the operator being used was 'and'.

In the second scenario with 'print ((8==8))' or (6 != 6)), because 8==8 did evaluate to True, it never made a difference that 6 != 6 evaluates to false since the operator being used was 'or'. Had we used the 'and' operator, it would have definitely be evaluated to False.

In the third scenario with 'print(not(3<=1))', you realize that the operator 'not' negates the False value which 3<=1 returns.

We will now substitute the floats for integers and target False evaluations:

```
print((-0.2 > 1.4) and (0.8 < 3.1)) # One original expression is False
print((7.5 == 8.9) or (9.2 != 9.2)) # Both original expressions are False
print(not(-5.7 <= 0.3))             # The original expression is True
```

In the above example,

- 'and' ought to have at least a single False expression evaluating to False
- 'or' ought to have the two expressions evaluating to False
- 'not' ought to have its inner expression being True for the new expression to evaluate to False

If the above results are not very clear to you, we will shortly go over some truth tables that will bring you up to speed.

You can also use 'and', 'not' and 'or' in writing your compound statements:

```
not((-0.2 > 1.4) and ((0.8 < 3.1) or (0.1 == 0.1)))
```

Now, first, focus on the innermost expression: '(0.8 < 3.1)' or (0.1 ==0.1). This particular expression evaluates to True because the two mathematical statements are True.

We can now take the returned 'True' value and merge it with the next inner expression which is '(-0.2>1.4)' and '(true)'. This example in particular returns false because the Mathematical statement '-0.2 > 1.4' is False and obviously, '(False)'and '(True)' returns False.

We finally have the outer expression which is 'not(False)', that evaluates to True, so the last returned value is: (if we print the statement) 'True'.

The logical operators and, not, and, or thus evaluate expressions and return the Boolean values.

Truth Tables

There is definitely a ton of stuff to learn about the branch of mathematics known as logic. However, we can learn some of it selectively to develop our algorithmic thinking for programming.

Look at the truth tables below for the comparison operator ==, and each one of the logic operators 'and', 'not' and 'or'. While it is possible not be able to reason them out immediately, it will prove helpful to work to memorize them so that they quicken your decision-making process.

The truth table for ==

x	==	y	Returns
True	==	True	True
True	==	False	False
False	==	True	False
False	==	False	True

The truth table for 'and'

x	and	y	Returns
True	and	True	True
True	and	False	False
False	and	True	False
False	and	False	False

The truth table for 'or'

x	or	y	Returns
True	or	True	True
True	or	False	True
False	or	True	True
False	or	False	False

The truth table for 'not'

not x	Returns
not True	False
not False	True

Truth tables are logic based mathematical tables- common as they are, they tend to be useful to memorize or always keep in mind when creating instructions or algorithms in any form of computer programming.

Boolean Operators for 'Flow Control'

To control the stream and program outcomes in form of flow control statements, you can use a condition, and then a clause.

A condition essentially evaluates down to a Boolean value of True or False, and presents a point where the program makes a decision i.e. a condition informs you when something evaluates to False or True.

On the other hand, a clause is a block of code following the condition and dictates the program outcome. That is, it forms the 'do this' part of the whole construction- 'if y is True, then do this.'

Look at the block of code below that describes comparison operators working together with conditional statements to control python program flow:

```
if grade >= 65:           # Condition
    print("Passing grade")    # Clause

else:
    print("Failing grade")
```

This program evaluates whether every student's grade is failing or passing. When a student (for instance), gets a grade of 83, the first statement evaluates to True, and the print statement of 'passing grade' is triggered. In case a student registers grade 59, the initial statement evaluates to False and thus, the program goes ahead to implement the print statement attached to the 'else' expression: 'failing grade'.

Since each Python object can be evaluated to False or True, expert python programmers recommend against comparing values to False or True since it is less readable and tends to return an unanticipated Boolean. This means you should try to avoid using something like 'sammy== True' in your programs and alternatively, make a comparison of 'sammy' to some other non-Boolean value that returns a Boolean.

Boolean operators usually present conditions we can use to decide the ultimate program outcome via flow control statements.

With all that covered, I know you are ready to make some decisions!

Conditional Statements: If and Else Statements

Every programming language typically has conditional statements. When you have conditional statements, you have code that runs sometimes (not all the time), even though it depends on the conditions of the program at that particular time.

When you execute every program statement fully, moving from the top to bottom with every line implemented in an order, you are not really instructing the program to actually evaluate particular conditions. We can use conditional statements to help programs to actually determine whether specific conditions are really being met.

To gain a better understanding of conditional statements, look at the following examples where they (conditional statements) would be used:

- If the student gets over 70% on his test, report that his grade passed; if not, report that his grade fails.
- If Jane has money in her account, calculate interest; if she does not, charge the penalty fee.
- If he buys 8 mangoes or more, calculate a 5% discount; if he buys less, then don't.

When you evaluate conditions and assign code to run based on whether those conditions are met or not, you are essentially writing conditional code. In this section, I will take you through all you need to know about writing conditional statements in Python.

If Statement

Let us begin with the 'if' statement that will determine whether a statement is false or true, and run code only when the statement is true. Open a file in a plain text editor and write this code:

```
grade = 70

if grade >= 65:
   print("Passing grade")
```

With the above code, you have the variable 'grade' and will give it an integer value of 70; you will then use the 'if' statement to evaluate whether the variable grade is more or equal (>=) to 65 or not. If it fails to meet this condition, you are instructing the program to print the string 'passing grade'.

Now save the program with the name 'grade.py' then use the command 'python grade.py' to run it.

Note that in our case, the grade of 70 does not meet the condition of being more than or equal to 65, and so, when you run the program, you will get the input below:

Output
Passing grade

We will not try to change the result of this program by altering the value of the variable 'grade' to 60.

```
grade = 60

if grade >= 65:
   print("Passing grade")
```

When you save and run the code, you will not receive any output because the condition was not met and you did not instruct the program to implement another statement.

Let us look at another example:

We will calculate whether or not a bank balance is less than 0. We will create a file named account.py and write the program below:

account.py

balance = -5

if balance < 0:
    print("Balance is below 0, add funds now or you will be charged a penalty.")

When you run the program attached with 'python account.py', you will get the following output:

Output
Balance is below 0, add funds now or you will be charged a penalty.

In the program you initialized, the 'balance' variable with the value '-5' is below 0. The balance did meet the 'if' statement condition (balance<0) and thus, when you save and run the code, you will get the string output. Again, if you alter the balance to 0 or some positive number, you will not get any output.

Else Statement

Even as an 'if' statement evaluates to false, it is possible that you will want the program to do something. In the grade example, you will want an output whether the grade is failing or passing. To do this, let us add an 'else' statement to the above grade condition, which we construct this way:

grade.py

```
grade = 60

if grade >= 65:
   print("Passing grade")

else:
   print("Failing grade")
```

The grade variable above contains a value of 60- thus, the 'if' statement will evaluate as false- and the program will not print the 'passing grade'. The subsequent 'else' statement will tell the program to do something anyway.

When you save and run, you will get the following output:

Output
Failing grade

By rewriting the program to give out the grade and a value of 65 or more, you will instead get the output 'passing grade.'

Rewrite the code as shown below to add an 'else' statement to the example on bank account:

account.py

```
balance = 522

if balance < 0:
   print("Balance is below 0, add funds now or you will be charged a penalty.")

else:
   print("Your balance is 0 or above.")
Output
Your balance is 0 or above.
```

Here, we changed the value of the 'balance' variable to a positive number in order for the 'else' statement to print. You can rewrite the value to a negative number to get the first 'if' statement to print.

When you combine the 'if' and 'else' statements, you will be constructing a conditional statement with two parts that will instruct your computer to implement a particular code whether the 'if' condition is met or not.

The else if statement

We have so far learnt a Boolean option for various conditional statements as each statement evaluates to either false or true. In most cases, you will want your program to evaluate more than two possible outcomes. For this, we will proceed to use the else if statement, which in Python, is written as 'elif'. The else if or 'elif' statement resembles the 'if' statement and evaluates another condition.

In our program on bank account, we may want to have some three distinct outputs for three situations that include the following:

- The balance is less than 0
- The balance equals to 0
- The balance is more than 0

The 'elif' statement goes between the 'if' statement and the 'else' statement as is shown below:

account.py

```
...
if balance < 0:
    print("Balance is below 0, add funds now or you will be charged a penalty.")

elif balance == 0:
    print("Balance is equal to 0, add funds soon.")

else:
    print("Your balance is 0 or above.")
```

Now we have three possible outputs you can receive once you run the program:

- If the 'balance' variable is equal to 0, you will then get the output using the 'elif' statement; this means the balance is 0 and you should add more money to your account.

- If the 'balance' variable is set to a positive number, you will get the output from the else statement; this means that your balance is either 0 or more.

- If the 'balance' variable is fixed to a number that is negative, the output will have to be the string from the 'if' statement; this means that the balance is less than 0 and therefore, you should add funds now or face a penalty.

What about an instance where you want to have over 3 options? This is possible by simply writing more than a single elif statement into your code. In the grade.py program, you can rewrite the code so that you will have a few letter grades agreeing with the ranges of numerical grades.

90 or more is equal to grade A

From 80 to 89 is equal to grade B

From 70 to 79 is equal to grade C

From 65 to 69 is equal to grade D

64 or below is equal to grade F

To run this code, we will require a single 'if' statement, 3 elif statements, as well an else statement to help you handle all the failing cases.

We will now rewrite the code from the above example to have strings printing out every letter grade. We can also keep the 'else' statement the same.

grade.py

```
...
if grade >= 90:
    print("A grade")

elif grade >=80:
    print("B grade")

elif grade >=70:
    print("C grade")

elif grade >= 65:
    print("D grade")

else:
    print("Failing grade")
```

The elif statements evaluate in order and thus, you can keep your statements basic. The program is completing the following steps:

- If the grade is more than 90, the program will print grade A; if it is less than 90, the program continues to the following statement...

- If the grade is more than, or equivalent to 80, the program will print grade B; if it is 79 or less, the program continues to the following statement...

- If the grade is more than, or equivalent to 70, the program will print grade C; if it is 69 or less, the program continues to the following statement...

- If the grade is more than or equivalent to 65, the program will print grade D; if it is 64 or less, the program continues to the following statement...

- The program prints 'failing grade' when all the conditions above are not met.

Nested if Statements

When you have started feeling comfortable with the elif, else, and if statements, you can start experimenting with the nested conditional statements. You can use the nested if statements when you want to check for secondary condition if the initial condition implements as true. In this case, you can include an if-else statement right in the middle on another if-else statement. Let us discuss a nested if statement:

```
if statement1:         #outer if statement
    print("true")

    if nested_statement:   #nested if statement
        print("yes")

    else:              #nested else statement
        print("no")

else:                  #outer else statement
    print("false")
```

Some possible outputs resulting from this code would be:

If the first statement (or statement 1) evaluates to true, the program evaluates whether the nested_statement is also evaluating to true. If both are true, you will get the following output:

Output
true
yes

However, when the first statement evaluates to true but the nested_statement evaluates to false, the output you get is:

Output
true
no

If the first statement evaluates to false, it means the if-else statement will not run and thus, the else statement runs alone. In such a case, the output you get is: *false*

You can also have many if statements nested all through your code:

```
if statement1:              #outer if
    print("hello world")

    if nested_statement1:       #first nested if
        print("yes")

    elif nested_statement2:     #first nested elif
        print("maybe")

    else:                   #first nested else
        print("no")

elif statement2:            #outer elif
    print("hello galaxy")

    if nested_statement3:       #second nested if
        print("yes")

    elif nested_statement4:     #second nested elif
        print("maybe")

    else:                   #second nested else
        print("no")

else:                       #outer else
    statement("hello universe")
```

In the above code, there is a nested if statement within every 'if' statement along with the elif statement. This allows for more options inside each condition.

You can look at an example of nested if statement using your grade.py program. You can check for whether or not a grade is passing first (more than or equivalent to 65%), and then look for the letter grade the numerical grade should be equal to. However, if the grade is not passing, you do not need to run through the letter grades. Instead, you can have the program reporting that the grade is failing.

The modified code with the nested if statement will appear like this:

```
...
if grade >= 65:
    print("Passing grade of:")

    if grade >= 90:
        print("A")

    elif grade >=80:
        print("B")

    elif grade >=70:
        print("C")

    elif grade >= 65:
        print("D")
else:
    print("Failing grade")
```

If you run the code with the 'grade' variable set to the integer value of 92, the first condition will be met and your program will print 'passing grade of:'. After that, it checks to find out whether the grade is more than, or equivalent to 90 and since this condition too is met, it prints out A.

If you run the code with the variable 'grade' set to 60, and then the first condition fails to be met, the program skips the nested if statements and moves on to the 'else' statement, and the program prints out 'failing grade'.

You can definitely add more options to this, and use another layer of nested if statements. Maybe you will want to evaluate for A-, A and A+ grades individually. We can also achieve that by first trying to see whether the grade is passing, then trying to check whether the grade is 90 or more, then checking whether (for instance) the grade is beyond 96 for an A+.

grade.py
```
...
if grade >= 65:
   print("Passing grade of:")

   if grade >= 90:
      if grade > 96:
         print("A+")

      elif grade > 93 and grade <= 96:
         print("A")

      elif grade >= 90:
         print("A-")
...
```

The code above shows that for variable 'grade' set to 96, the program runs the following:

- Check whether the grade is more than, or equivalent to 65; this is true.

- Print out the 'passing grade of:'

- Check whether the grade is more than or equal to 90; this is true

- Check whether the grade is more than 96; this is false

- Check whether the grade is more than 93 and also less than, or equivalent to 96; this is true

- Print A

- Drop these nested conditional statements and move on with the rest of the code

The program output for a grade of 96 will look like this:

Output
Passing grade of:
A

The nested if statements can give the chance to add a number of particular levels of conditions to your code.

When you use conditional statements such as the 'if' statement, you gain more control over what your program executes. Conditional statements instruct your program to evaluate whether or not a particular condition is being met. If the condition is met, it executes certain code, but if not, the program continues to move down to some other code.

In the next section, we will look at conditional statements that repeat themselves; the repeated implementation of code is mostly founded on Boolean conditions, so nothing new.

Chapter 11: Constructing 'While' Loops In Python

Computer programs come in handy in situations where you want to automate and repeat tasks because in so doing, you won't have to keep doing them manually. A great way to repeat the same tasks is by using loops and in this section of the book, we will be looking at the 'while' loop in Python.

The Boolean condition is the basis on which the 'while' loop executes the repeated implementation of code. The code that contains a 'while' block executes, so long as the 'while' statement is 'true'.

You can think of the 'while' loop as a conditional statement that repeats itself. The program executes the code after the 'if' statement. However, in the 'while' loop, the program goes to the start of the 'while' statement until it finds the condition as 'false'.

The while loops are conditionally based. This is not the same thing when it comes to 'for loops', which usually execute a specific number of times. Therefore, you do not need to know the number of times that you should repeat the code going in.

While loop

Here is how we construct while loops in Python:

```
while [a condition is True]:
    [do something]
```

What is being done in the program will continue to execute until the condition that is under assessment is no longer true.

Let us now try to create a little program for executing the 'while' loop. For this program, we will instruct the user to put in a password. As we go through the loop, there are 2 possible outcomes:

- If the password is right, the while loop exits
- If the password is not right, the while loop continues executing

We will build a file with the name password.py in any editor we choose, and start by initializing the 'password' variable as an empty string:

password.py

```
password = ''
```

You will use the empty string to take in input from the user in the while loop. We will now construct the 'while' statement together with its condition.

password.py
```
password = ''

while password != 'password':
```

In this case, the 'while' is followed by the 'password' variable. You are checking to see whether the variable is set to the 'password' string, but you can choose any string you would want. Thus, if the user inputs the 'password' string, then loop will cease and the program continues executing any code outside the loop. Nonetheless, if the string the user inputs is not equivalent to the 'password' string, the loop continues. After that, you will add the block of code that performs something inside 'while' loop:

password.py
```
password = ''

while password != 'password':
    print('What is the password?')
    password = input()
```

Within the 'while' loop, the program will run a print statement, which then prompts for the passwords. Afterwards, the 'password' variable is set to the input of the user via the 'input()' function.

The program checks to see whether the 'password' variable is assigned to the 'password' string; if it is, the 'while' loops ends automatically. Now give the program an extra line of code for when that occurs:

password.py
```
password = ''

while password != 'password':
    print('What is the password?')
    password = input()

print('Yes, the password is ' + password + '. You may enter.')
```

The final statement 'print()' is outside the 'while' loop; thus, when the user inputs 'password' as the password, that user will see the last print statement outside the loop. Nonetheless, if the user does not enter the 'password' word at all, he or she will not get to the final 'print()' statement and will thus be stuck in the infinite loop. What is the infinite loop really?

An infinite loop will occur when a program keeps executing within a single loop, at no time leaving it. You can simply press CTRL+C on the command line to exit from the infinite loops.

Now save the program and run: python password.py

You will get a prompt for a password, and may then test it with the different possible inputs. Look at the sample output from the program:

```
Output
What is the password?
hello
What is the password?
sammy
What is the password?
PASSWORD
What is the password?
password
Yes, the password is password. You may enter.
```

You should remember that strings are case sensitive unless you utilize a string function to change the string to all lower case (for instance) before you check.

A 'while' loop's example program

Now that we are making good progress with the general while loop's premise, we can create a guessing game based on command line, which makes effective use of the while loop. To understand how this program functions, ensure you read and understand the area on conditional statements well.

First, we will start by creating a file with the name 'guess.py' in your preferred text editor. You want your computer to come up with random numbers for the user to guess, and therefore, you will <u>import</u> the random module containing an import statement. If you are not familiar with this package, you can learn a lot about <u>creating random numbers</u> from Python docs.

guess.py

import random

After that, you will assign to the 'number' variable a random integer, making sure to keep it within the range of 1 through 25, which is inclusive. I hope this will not make the game too hard.

import random

number = random.randint(1, 25)

Here, you can get into your 'while' loop by first initializing a variable and then generating the loop.

```
guess.py
import random

number = random.randint(1, 25)

number_of_guesses = 0

while number_of_guesses < 5:
    print('Guess a number between 1 and 25:')

    guess = input()
    guess = int(guess)

    number_of_guesses = number_of_guesses + 1

    if guess == number:
        break
```

You have initialized the 'number_of_guesses' variable; this means that you increase it with every iteration of your loop so that you will not have an infinite loop. You then added the 'while' statement so that the variable 'number_of_guesses' is restricted to 5 in total. After making the fifth guess, the user will go back to the command line, and at this moment, if the user inputs something apart from an integer, he or she will get an error.

We added a 'print()' statement within the loop to prompt the user to input a number, which we took in with the function 'input ()' and set to the variable 'guess'. We then converted 'guess' to an integer, from a string.

Just before the loop is over, you also want to raise the 'number_of_guesses' variable by 1 so that you can iterate five times through the loop.

Lastly, you should write a conditional 'if' statement to check whether the 'guess' the user made is equal to the 'number' the computer made, and if so, you will use a <u>break statement</u> to exit the loop.

At this point, the program should be fully functional, and you can run it with this command: python guess.py

Even though it works, the user (at the moment) will not know whether their guess is correct and can even guess the entire five times without knowing whether they got it right. When you sample the current program's output, you get something like this:

Output
Guess a number between 1 and 25:
11
Guess a number between 1 and 25:
19
Guess a number between 1 and 25:
22
Guess a number between 1 and 25:
3
Guess a number between 1 and 25:
8

We will now add various conditional statements that are outside the loop to help the user to get feedback regarding whether or not he/she has guessed the number correctly. These go at the end of the current file:

```
guess.py
import random

number = random.randint(1, 25)

number_of_guesses = 0

while number_of_guesses < 5:
    print('Guess a number between 1 and 25:')
    guess = input()
    guess = int(guess)

    number_of_guesses = number_of_guesses + 1

    if guess == number:
      break

if guess == number:
    print('You guessed the number in ' + str(number_of_guesses) + ' tries!')

else:
    print('You did not guess the number. The number was ' + str(number))
```

Now the program tells the user whether he/she has correctly guessed the number (i.e. whether correct or wrong), which may not necessarily occur until the close of the loop after the user has actually run out of guesses. We will add a couple more conditional statements into the while loop to give the user a bit of help along the way. These will tell the user if the number they chose was too high or too low. This is to ensure they are more likely to guess the right number. Add this right before the 'if guess == number' line

```
guess.py
import random

number = random.randint(1, 25)

number_of_guesses = 0

while number_of_guesses < 5:
    print('Guess a number between 1 and 25:')
    guess = input()
    guess = int(guess)

    number_of_guesses = number_of_guesses + 1

    if guess < number:
        print('Your guess is too low')

    if guess > number:
        print('Your guess is too high')

    if guess == number:
        break

if guess == number:
    print('You guessed the number in ' + str(number_of_guesses) + ' tries!')

else:
    print('You did not guess the number. The number was ' + str(number))
```

When you run the program once more with 'python guess.py', you will note that the user secures a more guided help in their guessing. Thus, if the number generated randomly is 12, and he/she guesses 18, the program will inform the users that the guess is too high, and that they should thus adjust their subsequent guess accordingly.

To improve the code, there is a whole lot more you can do including handling errors for when the user fails to enter an integer. In this example though, you can see a 'while' loop working in a short command line program.

Chapter 12: Constructing 'For Loops' In Python Programming

As I mentioned in the previous section, in programming, the purpose of using loops is to automate and repeat similar tasks many times. In this section, we will be looking at the 'for loop' in python.

A 'for loop' executes the repeated implementation of code according to the loop counter or variable. This means you can utilize 'for loops' especially when the number of iterations is known before you enter the loop, which is nothing like the 'while loops' that are largely conditionally based.

The For Loops

In Python, we construct for loops as follows:

```
for [iterating variable] in [sequence]:
    [do something]
```

What is being done exactly is implemented until the sequence is finished. We will now look at a 'for loop' that iterates through values in a range:

```
for i in range(0,5):
    print(i)
```

By running the program, you get an output that looks something like this:

0

1

2

3

4

'i' is set up by this for loops as its iterating variable, and the sequence occurs in the range of zero to five.

We print a single integer for very loop iteration within the loop. Bear in mind that in programming, it is typical to start at index 0, which is why although five numbers are pointed out, they range from 0-4.

You will generally note and use 'for loops' when your program needs to repeat a block of code a particular number of times.

Using Range () In For Loops

One of Python's in-built types of immutable sequence is 'range ()'. Loops in particular uses 'range ()' to control the number of times the loop is repeated.

When you are working with range (), you can pass between one and three arguments of integers to it:

The value at which the integer starts is denoted by 'start'- if the value is not included, 'start' will start at 0.

The 'stop' integer is always essential and while counted up, it is not included.

'Step' sets how much increasing (or decreasing of negative numbers) the following iteration, - which if omitted, 'step' defaults to 1.

Let us look at a couple of examples of passing various arguments to 'range ()'. First of all, we will only pass the argument 'stop' so that the set up's sequence is 'range(stop)'.

```
for i in range(6):
    print(i)
```

In the above program, 6 is the 'stop' argument, and so the code iterates from 0-6-which is inclusive of 6. The output is:

0

1

2

3

4

5

Let's now highlight the 'range(start, stop)' with the specific values specified for when the iteration really starts and when they should stop.

```
for i in range(20,25):
    print(i)
```

The range here starts from 20 to 25- both of which are inclusive, so the output looks like this:

20

21

22

23

24

The range () step argument is the same as trying to specify stride while slicing strings; this means you can use it to skip values in the sequence. With the three arguments, 'step' will come in the last position: range(start, stop, step). First, let us use a 'step' that has a value that is positive.

for i in range(0,15,3):
　print(i)

Here, the for loop is strategically set up so that the numbers ranging from 0 to 15 print out, but a step of 3 so that it is only every third number that gets printed, like this:

0
3
6
9
12

You can also use a negative value for the 'step' argument so as to iterate backwards, but you will have to modify the 'stop' and 'start' arguments accordingly:

for i in range(100,0,-10):
　print(i)

In this case, 100 is the 'start' value, and 0 is the 'stop' value. -10 is the range; thus, the loop starts at 100 and ends at 0 while reducing by 10 with every iteration. You can see this happening in the output:

100
90
80
70
60
50
40
30
20
10

When you're using python to program, you will note that for loops usually utilize the range() sequence type as its iteration parameters.

Using Sequential Data Types In For Loops

Lists and other types of data sequence can also be used as iteration parameters in for loops. Instead of iterating through a 'range ()', you can easily define a list and iterate via it. Let us now assign a list to a variable and then try iterating through the list itself.

sharks = ['hammerhead', 'great white', 'dogfish', 'frilled', 'bullhead', 'requiem']

for shark in sharks:
   print(shark)

Here, we are printing out every item in the list. Even though we used the 'shark' variable, we could have called the variable any other valid name and we would not get any different output as you can see:

Output
hammerhead
great white
dogfish
frilled
bullhead
requiem

The above output clearly shows that the 'for' loop is iterated through the list, and prints every item from the list per line.

List and other sequence based data types such as tuples and strings are ideal to use with loops since they are iterable. You can bring together these datatypes with range () to take in items to a list- for instance:

In this case, I have included a placeholder string of 'shark' with every item of the 'sharks' list length. You can also use a 'for' loop to create a list from scratch.

integers = []

for i in range(10):
    integers.append(i)

print(integers)

In the example above, the list 'integers' gets initialized empty; nonetheless, the 'for' loop populates the list this way:

[0, 1, 2, 3, 4, 5, 6, 7, 8, 9]

Likewise, you can iterate through strings like this:

```
sammy = 'Sammy'

for letter in sammy:
  print(letter)
```

Output
S
a
m
m
y

You can iterate through tuples in the exact same format as iterating through strings or lists.

When you are iterating through a dictionary, it is good to keep the key, the value structure, in mind to make sure you are calling the right dictionary element. Look at this example that calls the key, and also the value:

```
sammy_shark = {'name': 'Sammy', 'animal': 'shark', 'color': 'blue', 'location': 'ocean'}

for key in sammy_shark:
  print(key + ': ' + sammy_shark[key])
```

Output
name: Sammy
animal: shark
location: ocean
color: blue

The iterating variable agrees with the dictionary keys whenever you use dictionaries with 'for' loops. On the other hand, the 'dictionary_variable (iterating variable)' usually agrees with the values. In the above case, the iterating 'key' variable was used to represent key, and 'sammy_shark (key)' was used to represent the values.

We often use loops to iterate and manipulate the sequential data types.

The Nested For Loops

We can nest loops in Python, as we can with other languages. A nested loop is a loop occurring within another loop, and is similar to the nested if statements in terms of structure. We construct them like this:

```
for [first iterating variable] in [outer loop]: # Outer loop
    [do something]  # Optional
    for [second iterating variable] in [nested loop]:   # Nested loop
        [do something]
```

The program will first meet the outer loop, implementing its first iteration. This first iteration activates the inner, nested loop that subsequently runs to completion. The program then returns to the top of the outer loop, to complete the second iteration and triggers the nested loop again. Once again, the nested loop runs to the end, and the program goes back to the top of the outer loop up until the sequence finishes or a break or some other statement interrupts the process.

Let us now implement a nested 'for' loop to help us understand more of this. In this case, the outer loop iterates through a list of integers known as 'num_list', and the inner loop iterates through strings list known as 'alpha_list'.

```python
num_list = [1, 2, 3]
alpha_list = ['a', 'b', 'c']

for number in num_list:
    print(number)
    for letter in alpha_list:
        print(letter)
```

Upon running the program, you will get this output:

Output
1
a
b
c
2
a
b
c
3
a
b
c

The output will show that the program finishes the outer loop's first iteration by painting 1, which then initiates completion of the inner loop, thus printing a, b, c successively. When the inner loop ends, the program will return to the top of the outer loop and print 2, and then print the inner loop entirely (a, b, c) and so on.

Nested for loops are necessary for iterating through items inside lists that comprise lists. If you use only one for loop, the program gives an output of every list as an item:

```
list_of_lists = [['hammerhead', 'great white', 'dogfish'],[0, 1, 2],[9.9, 8.8, 7.7]]

for list in list_of_lists:
    print(list)
Output
['hammerhead', 'great white', 'dogfish']
[0, 1, 2]
[9.9, 8.8, 7.7]
```

When you want to access every item of the internal lists, you will have to execute a nested 'for' loop.

```
list_of_lists = [['hammerhead', 'great white', 'dogfish'],[0, 1, 2],[9.9, 8.8, 7.7]]

for list in list_of_lists:
    for item in list:
        print(item)
Output
hammerhead
great white
dogfish
0
1
2
9.9
8.8
7.7
```

When you begin utilizing a nested 'for' loop, you become able to iterate over the individual items contained in the lists.

We have gone over how 'for' loops work in this section, and how to create them. For loops, continue looping through a block of code given a particular number of times.

Best for last perhaps... By now you should be looking for a way to make coding easier, and the next topic should help you achieve that—you will be able to group functions that belong together with classes, have a way to inherit other classes, and create a tree search structure for features in the linked classes. Overall, if you want to enjoy the ease of reusing present code as you write large programs, read on.

Chapter 13: Constructing Classes and Defining Objects

As a programming language, Python focuses on building reusable code patterns, which is why it is popularly considered object-oriented. Conversely, procedural programming is focused on explicit sequenced instructions. When you are working on complex programs, object oriented programming allows you to reuse and write code that is generally more readable, thus more maintainable.

Object oriented programming inconspicuously contains the distinction between objects and classes as one of its most important concepts:

- A class is a blueprint a programmer builds for an object. It defines the attributes that will characterize whichever object instantiated from this particular class.

- On the other hand, an object is an instance of a class. It is the realized class version; the class is exhibited in the program.

You will use these to build patterns (with respect to classes) and then utilize the patterns (with respect to objects). In this section, we will discuss about initializing attributes, instantiating objects and creation of classes with the use of the constructor method, and using multiple objects of the same class to work with.

Classes

You can think of classes as blueprints we use to create objects. We use the 'class' keyword to define classes, just as we use the 'def' keyword to define functions.

We will now define a class named 'shark' that contains two functions:

Swimming

Being awesome:

shark.py

```
class Shark:
  def swim(self):
    print("The shark is swimming.")

  def be_awesome(self):
    print("The shark is being awesome.")
```

Since these functions are indented within the class 'shark', we refer to them as methods (special functions defined within a class).

The word 'self' which is a reference to objects made based on this class, are the argument to these functions; 'self' will always be the initial parameter to reference instances, but it does not have to be the only one.

Defining this class then only created a shark object's pattern, which we will define later. It does not create any 'shark' objects. This means if you run your program beyond this stage, there will be nothing returned.

We therefore got a blueprint for an object by creating the 'shark' class above.

Objects

An object is just a class instance. You can take the 'shark' class, which we defined above then use it so as to create an object as well as its instance. We will now make a 'shark' object by the name 'sammy'- Sammy=shark().

In this case, by setting it equal to shark(), we initialized the object 'sammy' as a class instance. Now we will use both methods with the shark object 'sammy':

```
sammy = Shark()
sammy.swim()
sammy.be_awesome()
```

'sammy' is using both methods: be_awesome() and swim(). We called these using the dot operator (.) used to reference the object's attribute. In this case, the attribute is a method called with parentheses just as you would call with a function.

Since the 'self' keyword was a parameter of the methods as is defined in the class -'shark', the 'sammy' object is passed to the methods. The parameter-'self' makes sure the methods have a way of referring to attributes of the objects.

Nonetheless, when we call the methods, there is nothing passed within the parentheses and the 'sammy' object is automatically passed with the dot.

We will now add the object within the program's context:

shark.py

```python
class Shark:
    def swim(self):
        print("The shark is swimming.")

    def be_awesome(self):
        print("The shark is being awesome.")

def main():
    sammy = Shark()
    sammy.swim()
    sammy.be_awesome()

if __name__ == "__main__":
    main()
```

When you run the program:

The shark is swimming.
The shark is being awesome.

The 'sammy' object calls both methods in the program's function 'main()', thus making the methods run.

The Constructor

We use the constructor method to initialize data. The constructor runs when a class object is instantiated. It is also called _init_ method and will be the first class definition; it looks like this:

```python
class Shark:
    def __init__(self):
        print("This is the constructor method.")
```

If you add the _init_ method above to the 'shark' class in the above program, you would get an output that looks like the one below without doing any modification inside the 'sammy' instantiation:

This is the constructor method.
The shark is swimming.
The shark is being awesome.

The reason is that the constructor method is initialized automatically. You should carry out any initialization you want to do with your class objects with this method. Instead of using this constructor method, we can build one that uses a 'name' variable, which we can assign names to objects with. In this case, the 'name' will be passed as a parameter and we will set 'self.name' equivalent to 'name'.

shark.py

```
class Shark:
    def __init__(self, name):
        self.name = name
```

Next, we can modify the strings in our functions to reference the names as below:

shark.py

```python
class Shark:
    def __init__(self, name):
        self.name = name

    def swim(self):
        # Reference the name
        print(self.name + " is swimming.")

    def be_awesome(self):
        # Reference the name
        print(self.name + " is being awesome.")
```

Ultimately, you can set the 'shark' object's name 'sammy' as equal to 'Sammy' by passing it as a parameter of 'Shark' class.

shark.py

```python
class Shark:
    def __init__(self, name):
        self.name = name

    def swim(self):
        print(self.name + " is swimming.")

    def be_awesome(self):
        print(self.name + " is being awesome.")

def main():
    # Set name of Shark object
    sammy = Shark("Sammy")
    sammy.swim()
    sammy.be_awesome()

if __name__ == "__main__":
    main()
```

Now run the program: python shark.py

Sammy is swimming.
Sammy is being awesome.

As you can see, the name we passed to the object is printing out. We defined the _init_ method with the name of the parameter (together with the 'self' keyword) and also defined a variable in the method.

Since the constructor method is initialized automatically, we do not need to call it explicitly but pass the arguments within the parentheses that follow the class name when we make a new instance of the class.

We could also add another parameter like age if we wanted to by passing it to the _init_ method too.

```
class Shark:
    def __init__(self, name, age):
        self.name = name
        self.age = age
```

After that, when we build our object 'sammy', we can also pass the age of Sammy in the statement:

sammy = Shark("Sammy", 5)

We would also need to create a method in the class calling for age in order to make use of it. The constructor method enables us to initialize particular object attributes.

When Working With Multiple Objects

Classes are important because they let you create multiple similar objects on the same blueprint. For a better insight on how this works, we can add another 'shark' object to the program:

shark.py

```
class Shark:
    def __init__(self, name):
        self.name = name

    def swim(self):
        print(self.name + " is swimming.")

    def be_awesome(self):
        print(self.name + " is being awesome.")

def main():
    sammy = Shark("Sammy")
    sammy.be_awesome()
    stevie = Shark("Stevie")
    stevie.swim()

if __name__ == "__main__":
    main()
```

We have created another 'shark' object named 'stevie' and passed to it the name 'stevie'. In this case, we utilized the be_awesome() method with sammy and also the swim() with 'stevie'.

Now run the program: python shark.py

Sammy is being awesome.
Stevie is swimming.

The output shows that we are indeed using two different objects: the stevie and 'sammy' objects both of the 'shark' class. Classes allow you to make more than a single object following a similar pattern without building any of them from scratch.

Conclusion

That has been your guide to understanding programming with python. The book contains a few sections but is very deep. This means that in just a few days, you should be familiar with everything discussed herein, and be able to execute the projects discussed in it comfortably.

You should also note that we have covered a lot but not everything in python programming. Therefore, now that you have finished reading this book, it would be very wise of you to conduct more research on more topics to expand your knowledge of Python and programming as a whole.

Did You Enjoy This Book?

I want to thank you for purchasing and reading this book. I really hope you got a lot out of it.

Can I ask a quick favor though?

If you enjoyed this book, I would really appreciate it if you could leave me a positive review on Amazon.

I love getting feedback from my customers and reviews on Amazon really do make a difference. I read all my reviews and would really appreciate your thoughts.

Thanks so much.

P.S. You can click here to go directly to the book on Amazon and leave your review.

ALL RIGHTS RESERVED. No part of this publication may be reproduced or transmitted in any form whatsoever, electronic, or mechanical, including photocopying, recording, or by any informational storage or retrieval system without express written, dated and signed permission from the author.

Made in the USA
Middletown, DE
11 December 2017